# MATHEMATICS

# RAINFALL

*GET WET IN THIS RAINFALL*

LETS

# VOLUME EIGHT

*OBJECTIVES*

# TEMITOPE JAMES
## *Author and mathematician*

# ACKNOWLEDGEMENT

*I want to appreciate my loved ones for showing me care, love, support and affection towards the publication of this book. You all are wonderful and precious to me. May the lord bless all of you and keep you alive for me.*

*I also want to appreciate the students, the teachers, the schools and every individual that purchase this book and make use of it. Thanks for your cooperation to excellence, thanks for believing in the Flavor of mathematics and lastly, thanks for endeavoring to make use of the book. Success in your academic and physical life is guaranteed to you all. Amen*

# DEDICATION

*I dedicate this book to the Almighty God, for giving me the wisdom and knowledge to write this book with ease.*

*Thanks to my loving parents for always showing me love and care on the publication of this book. I also dedicate this book to my lovely daughter (Esther James) and my son (Flavor James). Your smile on your faces when you stir at me gives me joy to always appreciate your presence in my life.*

# PREFACE

The **FLAVOR OF MATHEMATICS** is an international practical workbook that covers the world mathematics curriculum. It is a book that is designed to inculcate and infuse practical hard work of solving various questions in mathematics which will help the students SUCCEED in both internal and external examinations in the world. The **FLAVOR OF MATHEMATICS** is good for learners, high school students, colleges, universities, job interview practical test and all forms of business oriented practitioner examinations for reference purposes of learning and infusing of mathematical studies into the lives of the students. Increasing the intellectual quotient of every user of this practical workbook is our duty and we have delivered this write up to boost your SUCCESS in mathematics.

This workbook is written, arranged, and designed for the use of reference studies for students who are willing to be sound, brave and psychologically strong in mathematics. This workbook covers important topics in mathematics and it is a book that is filled with the word of wisdom from great individuals and personalities in the world to create obedient and responsible students that will affect the society and there environment positively.

Therefore, the understanding of mathematics lies in the hands of students since this workbook is in existence. In fact, **THIS BOOK IS A PATHWAY TO SUCCESS THAT BREAKS THE YOKE OF FAILURE IN MATHEMATICS.** As it is said earlier, it lies in your hands if you are willing to break free from the oppression of mathematics by solving *every technical question in this workbook. This is a book that contains Over 1,500 Simple Basic Mathematics Questions from different 29 Topics. Have it in mind that 100% SUCCESS is guaranteed to every student that devout their time for study with this practical workbook. My advice for you is to have a copy of this book, sit down, eat, relax, bring out your pen, exercise book and calculator, and solve all questions in order for you to enjoy the* **FLAVOR OF MATHEMATICS.**

# *FLAVOR OF MATHEMATICS*

*M = Many people dislike me because they feel I am too difficult*

*A = All shall be incomplete without me*

*T = Try to practice me and you shall get used to me*

*H = How sad some people feel when they hear of me*

*E = Employ me and find out that I am unique among all other courses*

*M = Many set solutions to their mathematical problems through me*

*A = At least, I help those who work with me*

*T = Try me and you shall be great among equals*

*I = It will be good for you if you concentrate on me*

*C = Come to me and you will be good in all calculations*

# *The Flavor Of Mathematics*
## *INTRODUCTION*

*You are welcome to the Objective section of the MATHEMATICS RAINFALL from the platform of the Flavor of mathematics. The Flavor of mathematics is an international mathematical company that produces international practical workbook/textbook/journals which covers important basic mathematics curriculum to bring out the strength in YOU to solve mathematics vigorously without fear of failing mathematics in your INTERNAL and EXTERNAL examinations. In fact, the Flavor Of mathematics (volume two) is a book that will scare away laziness in the study of mathematics ONLY if you could attach yourself to the study of this book. It contains well composed Mathematics questions without options meant to foster the study of mathematics in your academic career. It is a book that will expose you to the act of looking beyond options in your examinations to prepare you for a guaranteed SUCCESS in your mathematical examinations.*

*This is also a book which will bring out the FLAVOR of your academic performance in mathematics and therefore; all students should have a copy of the MATHEMATICS RAINFALL from the platform of the Flavor Of Mathematics. We are concerned about your performance in mathematics and as a mathematical brand company; this book is published to bring out the best in you for you to EXCEL in mathematics without the fear of options for required answers.*

*Please sit back, relax and SOLVE all questions with passion for your SUCCESS in mathematics. Have it in mind that as you are solving the questions in this book, imagine you are in the examination hall because all questions here are of international standard which covers all basic mathematics curriculum of the world. The FLAVOR OF MATHEMATICS is a company that will bring out the BEST in your career to excel in mathematics.*

*Therefore; the FLAVOR OF MATHEMATICS is your book to behold for SUCCESS in theoretical mathematics. From the team of the FLAVOR OF MATHEMATICS, we wish you good luck and hard Work to solve ALL QUESTIONS conveniently.*

# *Welcome to the objective platform*
## *Of the*
## *MATHEMATICS RAINFALL*

*Welcome to the volume eight section of the mathematical rainfall from the platform of the FLAVOR OF MATHEMATICS. You are advice to concentrate on all the questions and solve them without the assistance of anybody. Do not ask questions, do not be lazy, do not sleep, do not weary, do not be afraid, do not give room for NEGATIVE side attraction solving these questions. Just Sit, relax, eat and pick up your pen or device to solve the following below;*

## LOGARITHMS

*Logarithms under this chapter are based on both the logarithms of base 10 and the theory of logarithms. Both the normal logarithm and the theory of logarithm uses the addition, division, multiplication and subtraction but only the difference is that, the theory of logarithm does NOT use the ANTILOGARITHM for calculation. Sit, eat, relax, concentrate and enjoy the FLAVOR OF MATHEMATICS.*

1. $\log_8 x = {}^1/_3$     (a) 1   (b) 2   (c) 3   (d) 4

2. $\log_{10}x + \log_{10}x^3 + \log_{10}x^5 + \log_{10}x^6 + \log_{10}x^7 + \log_{10}x^9 + \log_{10}x^{18} = 49$
    (a) 5   (b) 7   (c) 10   (d) 15

3. $\log_{10}{}^{1/2} + \log_{10}{}^{1/5} - \log_{10}{}^{7/10}$     (a) 1   (b) 2   (c) 3   (d) 4

4. Simplify $\log_{10}27 / \log_{10}81$     (a) $\frac{3}{4}$   (b) ${}^4/_3$   (c) ${}^3/_2$   (d) ${}^2/_3$

5. $\log_{10}\sqrt{8} + \log_{10}\sqrt{10} - \log_{10}\sqrt{8}$     (a) 1   (b) $\frac{1}{2}$   (c) ${}^3/_2$   (d) ${}^3/_4$

6. $\log_{10}(2x + 4) - \log_{10} 4 = 1$     (a) 7   (b) 240   (c) ${}^{21}/_{23}$   (d) 18

7. $\log_2 4$     (a) 1   (b) ${}^1/_5$   (c) 2   (d) $\frac{1}{2}$

8. $\log_4 x - \log_4(x - 2) = 3$, express x in term of y
    (a) ${}^{128}/_{63}$ (b) ${}^9/_{125}$ (c) ${}^{63}/_{128}$ (d) none

9. $\log\sqrt{49} - \log\sqrt{7}$     (a) 1   (b) 3   (c) 2   (d) 4

10. $2\log_3 3 - 3\log_3 3 + 4\log_3 3$     (a) 1   (b) 2   (c) 3   (d) 4

11. Find x in terms of y; $\log_4 x + 3\log_4 y = 3$
    (a) $({}^4/_y)^{1/3}$ (b) $({}^y/_4)^{1/2}$ (c) $({}^y/_4)^{-1/3}$ (d) $({}^4/_y)^3$

12. Solve $2\log_2 4 + 2\log_2 2$     (a) 1   (b) 3   (c) 2   (d) 6

13. $\log_3 27$     (a) 1   (b) 2   (c) 3   (d) 4

14. $\log_{10} 10000$     (a) 1   (b) 2   (c) 3   (d) 4

15. $\log_3 27 - \log_3 81$     (a) ${}^4/_3$   (b) $\frac{3}{4}$   (c) $\frac{1}{4}$   (d) ${}^4/_5$

16. $\log_3 3\sqrt{9}$     (a) 2   (b) $\frac{3}{4}$   (c) $\frac{1}{4}$   (d) ${}^4/_3$

17. Log $_2$ ½ = x; find x        (a) 1   (b) 2   (c) – 1   (d) – 2

18. Log$_8$x = 2        (a) 64   (b) $2^3$   (c) 16   (d) 32

19. Log$_{64}$x = $^1/_3$        (a) 1    (b) $^1/_4$   (c) 3    (d) 4

20. Log$_4$x = – 2        (a) $^1/_{16}$   (b) $^1/_8$   (c) ¼   (d) ½

21. Log$_3$³ + log$_3$²⁷        (a) 1   (b) 2   (c) 3   (d) 4

22. Log a$^x$ – Log a$^y$        (a) $^x/_y$   (b) $^y/_x$   (c) $x^2/_y$   (d) $y/x^2$

23. Log$_2$8        (a) 2    (b) 3   (c) ¼   (d) $^4/_5$

24. Find the reciprocal of Log$_9$81        (a) 2   (b) ½   (c) 2½   (d) $^3/_2$

25. Log $(^3\sqrt{8}/_{\log 8})$        (a) $^1/_3$   (b) 3    (c) $^3/_2$   (d) $^2/_3$

26. Log$_{10}$ 0.01        (a) 1   (b) – 2   (c) 3    (d) 4

27. Log$_2$ $^3/_2$ + Log$_2$ $^{16}/_{27}$ – Log$_2$ $^1/_9$        (a) 1   (b) 2   (c) 3   (d) 4

28. Log$_9$⁷²⁹ – Log$_9$⁹        (a) 1   (b) 2   (c) 3   (d) 4

29. Simplify $^{\text{Log }\sqrt{49}}/_{\text{Log }7}$        (a) 2   (b) ½   (c) 7   (d) 1

30. Simplify $^{\log 243}/_{\log 2187}$        (a) $^5/_7$   (b) $^3/_7$   (c) $^1/_7$   (d) $^2/_7$

31. Solve these by using Logarithm table $^{84600}/_{(25.33 \times \sqrt{124.3})}$

       (a) 4687   (b) 468.7   (c) 46.87   (d) 4.687

32. Simplify Log $_{10}$¹⁰⁰/ Log $_{10}$$^{\sqrt{10}}$        (a) 2   (b) 4   (c) 3   (d) 1

33. Solve these with Logarithm table $^{(38.86 \times 509.2)}/_{76.71}$
       (a) 2.579   (b) 257.9   (c) 25.79   (d) 2579

34. Solve Log$_3$³ + Log$_3$⁹        (a) 0    (b) 3    (c) 2   (d) – 3

35. Simplify Log$_2$⁴ + Log$_2$¹⁶ = x        (a) 4   (b) 5   (c) 6   (d) 8

36. $^{\text{Log }y}/_{\text{Log }\sqrt{y}}$        (a) 1   (b) 2   (c) 3   (b) 4

37. Log$_{10}$ (x$^2$ + 4) = 4 + Log$_{10}$x – Log$_{10}$100
       (a) 9.99, 0.05 (b) 99.9, 0.09 (c) 9.9, 5 (d) 9, 0.5

38. Log$_u$8 + Log$_u$4 = 8        (a) 2 (b) $^{-1}/_2$ (c) $^3/_2$ (d) none of the above

39. 4Log (t – 2) = 2Log4        (a) 1   (b) 2   (c) 3   (d) 4

40. Log$_4$ $(^{38}/_{19})$ + 2Log$_4$ $(^2/_4)$ – Log$_4$ ⁹
       (a) Log$_4$⁶ (b) Log$_4$ $^1/_6$ (c) Log$_4$² (d) Log$_4$ $^3/_2$

41. Solve with Logarithms table $^{(193.4 \times 28.9)}/_{48.2}$
       (a) 4315   (b) 43.15   (c) 431.5   (d) 4.315

42. Using Logarithm table, what is 0.6924 × 0.0837?
       (a) 56    (b) 0.56   (c) 5.6   (d) 0.056

43. In Logarithm table 89.95 ÷ 0.0025
       (a) 3593   (b) 3592   (c) 3591   (d) 3590

44. Log$_{10}$ (9x + 3) – Log$_{10}$ (4x – 1) = 1
       (a) $^{31}/_{18}$ (b) $^{13}/_{31}$ (c) $^{13}/_{33}$ (d) $^{33}/_{23}$

45. Log$_{10}$ (x$^2$ + 3) = 3 – Log$_{10}$x – Log$_{10}$²⁷
       (a) 36.91, 0.08 (b) 3.691, 0.8 (c) 369.1, 0.008 (d) 36, 8

46. Use the log table to solve $198.9^5/22.9$

        (a) 490. 49  (b) 49490 (c) 44904 (d) 49409

47. $\log_5 5/_{10} + \log_5 25 + \log_5 250$     (a) 4   (b) 3   (c) 2   (d) 1

48. $\log_4 4096 = y$     (a) 4   (b) 2   (c) 6   (d) 5

49. $\frac{1}{2}\log_3 \sqrt{9} = x$     (a) $^1/_2$   (b) $^4/_7$   (c) $^2/_5$  (d) ¾

50. $\log_{10} 1000 - \log_{10} 100$     (a) 1½   (b) $^3/_5$   (c) $^2/_3$   (d) ¾

51. $\log_7 {}^3\sqrt{7}$     (a) ¾   (b) $^2/_3$   (c) $^5/_7$   (d) 1½

52. $\log_4 64 - \log_4 8 + \log_4 2$     (a) 1   (b) 2   (c) 3   (d) 4

53. $3\log (^3/_5) + 2\log (^2/_3) - \log (^3/_2)$

        (a) $2\log {}^2/_5$  (b) $3\log {}^2/_5$  (c) $4\log {}^2/_5$  (d) $5\log {}^2/_5$


      Given that Log 13 = 1.1139, Log11 = 1.0414 and Log8 = 0.9030

54. What is Log169?

        (a) 2.0828  (b) 2.2278  (c) 1.69  (d) 1.806

55. What is Log121?     (a) 2.0828  (b) 1.69  (c) 2.2278  (d) 1.806

56. What is Log 64?     (a) 2.0878  (b) 2.2278  (c) 1.806  (d) 1.2132

57. Solve Log 512 + log 121     (a) 0.3072  (b) 4.792  (c) 6.9541  (d) 2.32

58. Solve $\frac{1}{2}\log_2 4 - \log_2 x - {}^1/_2\log_2 9 = \log_2 4$

      (a) $\log_2 22$  (b) $\log_2 43$  (c) $\log_2 6$  (d) $\log_2 8$

59. $\log_{27} y = 0.333$     (a) 1   (b) 2   (c) 3   (d) 4

60. $2\log_2 {}^2/_3 + \log {}^{81}/_{16} + \log_2 {}^{64}/_9$     (a) 1   (b) 2   (c) 3   (d) 4

# ARITHMETIC AND GEOMETRIC PROGRESSION

*Arithmetic progression is shortening as A.P while the geometric progression is shortening as G.P. The A.P is a sequence that increases or decreases between two consecutive terms while the G.P is a sequence in which any two consecutive terms differ by a constant factor. Also, the solving of the series and sequences are also stated in this chapter. Sit, relax, concentrate and enjoy the FLAVOR OF MATHEMATICS.*

1. Find the 6$^{th}$ term of the A.P of 2, 5, 8 and 11
   (a) 4  (b) 15  (c) 16  (d) 17

2. Find the 4$^{th}$ term of the G.P of 4, 8 and 16
   (a) 23  (b) 32  (c) 16  (d) 28

3. Write down the 2$^{nd}$ and 3$^{rd}$ terms of the sequence even as $2n - (n^{-1})$
   (a) $^2/_7$, $^{17}/_3$   (b) $^7/_2$, $^{17}/_3$   (c) $^7/_2$, $^3/_{17}$  (d) $^7/_2$, $^{-13}/_7$

4. The n$^{th}$ term of a sequence is given as $(^1/_n)$. Find the first four terms
   (a) $^1/_2$, $^1/_3$, 1, ¼  (b) ¼, ½, $^1/_3$, 1 (c) ½, ¼, $^1/_3$, 1 (d) 1, ½, $^1/_3$, ¼

5. The common difference of the A. P – 3, 2 and 7 is
   (a) 2  (b) 3  (c) 4  (d) 5

6. Find the first term of a G.P when the common ratio is 2 and the 4$^{th}$ term of the G.P is 64.
   (a) 7  (b) 8  (c) 9  (d) 10

7. The sum of the three consecutive terms of an A.P is 18, if their product is 120, find the terms
   (a) 9, 12, 6 (a) – 9, 6, – 2, (a) 10, 6, 2 (a) – 6, – 2, 10

8. The 5$^{th}$ term of an A.P is 16 and the 4$^{th}$ term of an A.P is 22. Find the 8$^{th}$ term.
   (a) – 3  (b) – 2  (c) – 1  (d) 1

9. Find the sum of the series $n^3 + 2n$ from 1$^{st}$ term to the 3$^{rd}$ term in order.
   (a) 12 + 3 + 33  (b) 33 + 12 + 3 (c) 3 + 33 + 12 (d) 3 + 12 + 33

10. Find the product of the 3$^{rd}$ and 4$^{th}$ term of the sequence $n + n^2$
    (a) 199   (b) 240   (c) 32   (d) 8

11. Write out the series $2n + 5n^2$ from 1$^{st}$ term to the 4$^{th}$ term in order.
(a) 7 + 88 + 51 + 24 (b) 7 + 24 + 51 + 88 (c) 7 + 51 + 88 + 24 (d) 51 + 88 + 7 + 24

12. Find the sum of A.P in the following; 6, 10, 14, and 18 using the 8$^{th}$ term.   (a) 120 (b) 140 (c) 160 (d) 180

13. Find the sum of the 5$^{th}$ terms of an A.P of the sequence 9, 12, 15 and 18.
    (a) 75 (b) 76 (c) 77 (d) 78

14. The common ratio of a G.P is 4. If the 6$^{th}$ term is equal to 5120, find the 2$^{nd}$ term
    (a) 17 (b) 20 (c) 23 (d) 32

15. The sum of the 4$^{th}$ term is 18 and the sum of the 8$^{th}$ term is 20. Find the product of the A. P of the 3$^{rd}$ & 4$^{th}$ term.
    (a) 10   (b) 11   (c) 12   (d) 13

16. The 4$^{th}$ term has the first term and common ratio as 4 and 2. Find the G.P
    (a) 12  (b) 22  (c) 42  (d) 32

17.  The addition of the 3$^{rd}$ & 5$^{th}$ term of an A.P is 12 and the 8$^{th}$ term is 28. Find the 17$^{th}$ term        (a) 7207   (b) 77.679   (c) 77.5   (d) 7.75

18.  The 4$^{th}$ term of an A.P is z & the 5$^{th}$ term of an A.P is w. Find the first term & the C.D
       (a) $d = z/w$, $a = 3w + z^2$ (b) $d = w - z$, $a = 4z - 3w$ (c) $d = w + z$, $a = 4z + 3w$

19.  What is the value of the 1$^{st}$ term when it has the C.D as 3 and the A.P is 24 with 5$^{th}$ term?        (a) 12   (b) 13   (c) 14   (d) 15

20.  The 4$^{th}$ term of a A.P is 12 more than the 2$^{nd}$ term, while the 8$^{th}$ term is 16. What is the product of the 1$^{st}$ term and the common difference?
       (a) 230   (b) -156   (c) 120   (d) 150

21.  Find the sum of A.P of 6$^{th}$ term of the sequence 7, 14, 21 and 28
       (a) 145   (b) 146   (c) 147   (d) 148

22.  Find the sum of infinity of the series 5, 1 and $^1/_5$.
       (a) 6¼   (b) 5$^1/_3$   (c) 4¼   (d) 3½

23.  What is the sum of the 14$^{th}$ term of an A.P whose first term is 4 & the C.D is 8?        (a) 748   (b) 478   (c) 784   (d) 874

24.  If 2, a, b, 250 is a G.P, find a and b
       (a) 10, 50   (b) -11, 52   (c) -9, 42   (d) 10, 58

25.  The work of a cleaner carries a salary of $1200 arising by annual increment of $80 to $1,760 per year. What is the total amount the cleaner would earn if he hold to the past for 20 years?
       (a) $33,900   (b) $32,690   (c) $33,920   (d) $32,960

26.  In an A.P, the differences between the 7$^{th}$ and the 4$^{th}$ term is 27 and the 6$^{th}$ term is three times the 3$^{rd}$ term. Find the common difference
       (a) 4   (b) 7   (c) 9   (d) 12

27.  The 3$^{rd}$ term of an A.P is 40 and the 4$^{th}$ term is 18 more than the 6$^{th}$ term. Find the 8$^{th}$ term of the sum of A.P.
       (a) 122   (b) 212   (c) 221   (d) 201

28.  In a G.P, the first term is 2 and the common difference is 3. Find the sum of the 6$^{th}$ term        (a) 728   (b) 287   (c) 782   (d) 872

29.  The sum of infinity is 28 and the first term is 3. Find the common ratio
       (a) $^{28}/_{25}$   (b) $^{25}/_{28}$   (c) $^{28}/_{29}$   (d) $^{29}/_{28}$

30. A geometric progression is also known as        (a) Quadratic equation (b) Exponential function (c) Linear sequence (d) Set

31.  The 4$^{th}$ term of an A.P is 10 more than the 6$^{th}$ terms while the 8$^{th}$ terms is one – half of 24. Find the sum of the 7$^{th}$ and 10$^{th}$ term of an A.P        (a) 11 (b) 41 (c) 67 (d) 40

32.  The 9$^{th}$ term of an A.P is 12 more than twice the 6$^{th}$ term while the 8$^{th}$ term of an A.P is 18 more than one-half the 4$^{th}$ term. Find the difference between the 16$^{th}$ and the 8$^{th}$ term if an A.P
       (a) 19.8 (b) 22.7   (c) 25.1   (d) 27.4

33.  The seventh term of an A.P is 24 more than the 3$^{rd}$ term while the 6$^{th}$ term of an A.P is 18 more than the 3$^{rd}$ term. Find the 18$^{th}$ term
       (a) 201 (b) 21 (c) 102 (d) 210

34.  How many terms have the A.P when the first term and the common

difference is 2 and 4 and the last term is 42?

(a) 9   (b) 10   (c) 11   (d) 12

35. The 4$^{th}$ term of a G.P is 324 while one third of the first term is 4. Find the common ratio   (a) 1   (b) 2   (c) 3   (d) 4

36. What is the sum of the 6$^{th}$ term of an A.P when the first term is 4 and the common difference is 8?   (a) 144   (b) 414   (c) 44   (d) 141

37. When the 4$^{th}$ term of an A.P is 4 more than the 3$^{rd}$ term, one – half of the product of the first term and common difference is 12. Find the first term and C.D

(a) a = 4, d = 6 (b) a = 2, d = 4 (c) a = 2, d = 12 (d) a = 8, d = 3

38. The sequence of $3^{3n-3}$ is 27. What is the n$^{th}$ term?

(a) 1   (b) 2   (c) 3   (d) 4

39. Write the sequence of the 3$^{rd}$ and 4$^{th}$ term of $1/n + n/4$.   (a) 3$^{rd}$ = $13/12$, 4$^{th}$ = $5/4$ (b) 4$^{th}$ = $-5/4$, 3$^{rd}$ = $-13/12$ (c) 3$^{rd}$ = $5/4$, 4$^{th}$ = $13/12$ (d) 3$^{rd}$ = $12/13$, 4$^{th}$ = $4/5$

40. The 4$^{th}$ term of an A.P is 48 while the 5$^{th}$ term is 64. Find the sum of the 2$^{nd}$ and 6$^{th}$ term   (a) 93   (b) 96   (c) 106   (d) 112

41. The 18$^{th}$ term of an A.P is 98, given that the common difference is 9, Find the first term and the 38$^{th}$ terms.   (a) a = 55, 38$^{th}$ = 728 (b) a = – 55, 38$^{th}$ = 872 (c) a = – 55, 38$^{th}$ = 278 (d) a = 55, 38$^{th}$ = – 872

42. The sum of the first seventh terms of the series $1/7 + 2/7 + 4/7 + $ ....is

(a) $27/7$   (b) $127/7$   (c) $43/7$   (d) $5/7$

43. The third term of a G.P is 245. When the first term is 5, find the common ratio   (a) 5   (b) 6   (c) 7   (d) 8

44. The sum of the 7$^{th}$ term of an A.P is 64 and the sum of 6$^{th}$ term of an A.P is 42. What is the difference in the 8$^{th}$ and 4$^{th}$ term of the sum of A.P?

(a) 79.42   (b) 97.4   (c) 47.9   (d) 49.7

45. Find the product of the 6$^{th}$ term of an A.P and the 8$^{th}$ term of an A.P when the first term is 4 and the common difference is 2

(a) 522   (b) 252   (c) 225   (d) 502

46. Find the 27$^{th}$ term of the sum of G.P when first term is 2 and common ratio is 3   (a) $3^{27}-1$   (b) $3^{26}$   (c) $6^{26}$   (d) $6^{27-3}$

47. What is the n$^{th}$ term of the G.P when the sequences are 5, $-5/2$, $5/4$, $-5/8$ ........ 40?   (a) – 2   (b) – 1   (c) 1   (d) 2

48. Find the 6$^{th}$ and 10$^{th}$ term of a G.P whose first term and common ratio are 6 and 8 respectively.

(a) 6$^{th}$ = 196608, 10$^{th}$ = 6(8)$^6$   (b) 6$^{th}$ =196612, 10$^{th}$ = 9$^{10}$

(c) 6$^{th}$ = 196608, 10$^{th}$ = 6 (8)$^9$   (d) 6$^{th}$ = 169866, 10$^{th}$ = 7(8)$^7$

49. The sum of infinity is 20 and the common ratio is 4. Find the first term

(a) 40   (b) 50   (c) – 60   (d) 70

50. The sum of infinity is 9 while the first term is 2. What is the common ratio?   (a) $9/7$   (b) $7/9$   (c) $1 1/3$   (d) $2 1/7$.

# POLYGONS

*A polygon is described as a closed figure in which at least three straight line is bounded together. Polygon is named according to the number of its sides. The solving includes finding its sides or given angles. Sit, concentrate and enjoy FLAVOR OF MATHEMATICS.*

1.  A regular polygon has an interior angle of $120^0$. The number of side it has is
(a) 2 sides  (b) 4 sides  (c) 6 sides  (d) 10 sides

2.  The interior angle of a polygon are z, $(z + 10)^0$, $(z + 20)^0$, $(z + 30)^0$ and $(z + 40)^0$. Find z          (a) $78^0$ (b) $88^0$ (c) $98^0$ (d) $108^0$

3.  An angle of heptagon is $60^0$. If the other angles are equal to each other, find the angle          (a) $120^0$  (b) $130^0$  (c) $140^0$ (d) $150^0$

4.  An angle of a heptagon is $120^0$. If the other angles are equal to each other, find the angle.
(a) $120^0$  (b) $130^0$  (c) $140^0$  (d) $150^0$

5.  Five angles of a decagon are equal and each of the other five is $60^0$ greater than each of the first five angles. Find the first five angles
(a) $121^0$  (b) $173^0$  (c) $114^0$  (d) $119^0$

6.  What is the 3<sup>rd</sup> angle of the triangle when the other two angles is $(x + 90)^0$ and $(3x + 60)$?
(a) $(30 – 4x )^0$  (b) $(30 + 4x)^0$  (c) $(4x – 28)^0$  (d) $(28 + 4x)^0$

7.  Four of the angles of an octagon are x each. The other four angles are 4x each. Find x          (a) $37^0$  (b) $42^0$  (c) $54^0$  (d) $58^0$

8.  Write down the number of degrees in each angle of a 18 sided polygon.
(a) $140^0$  (b) $150^0$ (c) $160^0$  (d) $70^0$

9.  When one of an angle of a hexagon is $170^0$, find each of the other angles when they are all equal to each other.
(a) $110^0$  (b) $120^0$  (c) $130^0$  (d) $140^0$

10. Find the seventh angle of a heptagon when each other angles are $130^0$ each          (a) $110^0$  (b) $120^0$  (c) $130^0$  (d) $140^0$

11.    $50^0$, $70^0$, $90^0$, $120^0$, and $150^0$ are five angles of a hexagon. Find the 6<sup>th</sup> angle          (a) $210^0$  (b) $220^0$  (c) $230^0$  (d) $240^0$.

12.    One angle of a Nonagon is $20^0$. Find each of the other angles if they are equal to each other.
(a) $140^0$  (b) $155^0$  (c) $1444^0$  (d) $162^0$

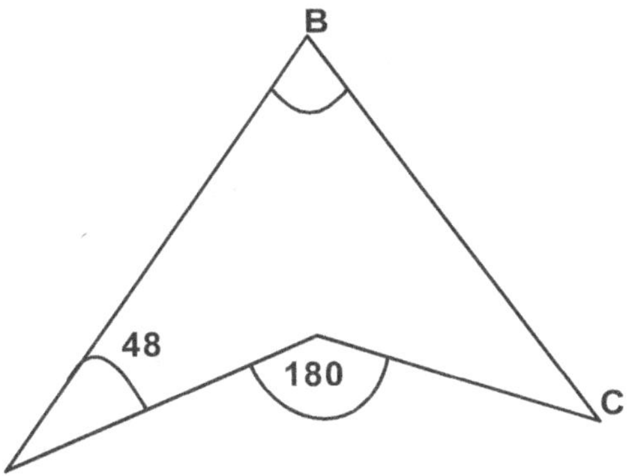

**13.**
When B is 90⁰, find C    (a) 91⁰   (b) 40⁰   (c) 42⁰  (d) 44⁰

**14.**    If three angles of a triangles are $(x + 42)^0$, $(2x + 58)^0$ and $(x - 90)^0$. Find x.                        (a) 40  (b) 43  (c) 42.5  (d) 41

**15.**    A regular polygon has an angle of 160⁰. How many sides does it has?
(a) 16 sides (b) 17 sides (c) 18sides (d) 19 sides

**16.**    If $(x + 92)^0$, $(2x - 60)^0$ and $(x + 48)^0$ are angles of quadrilateral, find the 4th angle
(a) $(280 - 4x)^0$   (b) $(160 + 4x)^0$   (c) $(100 - 2x)^0$    (d) $(100 - 4x)^0$

**17.**    A regular polygon has an angle of 90⁰. How many sides does it have?
(a) 6 sides  (b) 4 sides  (c) 3 sides  (d) 7sides

**18.**    If the angle of a nonagon are $y^0$, $(y + 10)^0$, $(y + 20)^0$, $(y + 30)^0$, $(y + 40)^0$, $(y + 50)^0$, $(y + 60)^0$, $(y + 70)^0$ and $(y + 80)^0$. Find y
(a) 60⁰   (b)  70⁰  (c)  80⁰ (d)   100⁰

**19.**    One angles of an heptagon is 320⁰. Find the angles given that they are equal to each other.   (a) 92⁰    (b) 97⁰ (c) 70ⁿ   (d) 120⁰.

**20.**    The angle of pentagon are $(x + 20)^0$, $(x - 60)^0$, $(x + 110)^0$, $(3x - 4)^0$ and $(x + 35)^0$. Find x
(a) 62.7⁰   (b) 67⁰   (c) 52.8⁰   (d) 75⁰

**21.**    What is x in a reflex diagram given such that its sides are x, 2x, 3x, 4x and 5x?                        (a) 24⁰   (b) 36⁰   (c)40⁰        (d) 50⁰

**22.**    In a regular polygon of x sides, each interior angles is 150⁰, what is x?
(a) 8 sides   (b) 11 sides      (c) 20sides    (d) 12 sides

**23.**   One interior angles of a convex hexagon is 430⁰ and each of theremaining angles are equal to x. find x.
(a) 40⁰  (b) 50⁰   (c) 58⁰  (d) 64⁰

**24.**  The sum of the angles of a regular polygon is 1440⁰. How many sides does the polygon have?
(a) 10Sides  (b) 12Sides  (c) 14Sides  (d) 18 Sides.

25. Find the size of an exterior angle of a regular convex polygon with 18 sides $\qquad$ (a) $10^0$ (b) $20^0$ (c) $30^0$ (d) $40^0$

26. A duodecagon has eleven of its angles equal to each other. If the 12th angle is $139^0$. Find the size of each of the other angles.
(a) $132^0$ (b) $142^0$ (c) $151^0$ (d) $173^0$.

27. Each Interior angles of a regular hexagon is given as
(a) $110^0$ (b)$120^0$ (c) $130^0$ (d) $140^0$

28. What is the value of x when the interior angle of an heptagon is given; $68^0$, $(x + 28)^0$, $(x - 20)^0$, $49^0$, $(3x + 58)^0$, $(x + 37)^0$ and $(2x + 80)^0$ ?
(a) $75^0$ (b) $85^0$ (c) $95^0$ (d) $105^0$

29. If the angles of quadrilateral are $(x + 10)^0$, $(2x + 40)^0$, $(3x + 90)^0$ and $2x$. Find x (a) 225 (b) 27.5 (c) 32.75 (d) 38.7

30. What is the side of an exterior angle of a regular undecagon?
(a) 32. 7 (b) 38.72 (c) 36.9 (d) 37.5

31. What is the size of an exterior angle of a regular duodecagon?
(a)$10^0$ (b) $20^0$ (c) $30^0$ (d) $40^0$

32. A 13 sided polygon is known as
(a) Duodecagon (b) Triodecagon (c) Decagon (d) Undecagon

33. One angles of a pentagon is $140^0$. Find each of the other angles, given that they are all equal to each other.
(a) $100^0$ (b) $110^0$ (c) $120^0$ (d) $130^0$

34. Write down the number of degrees in each angle of a regular 16 side polygon. (a) $150.5^0$ (b) $157.5^0$ (c) $166.5^0$ (d) $167.5^0$

35. The interior angle of a pentagon is $(x - 10)^0$, $(x + 70)^0$, $(x + 30)^0$, $(x + 20)^0$ and $(x + 15)^0$. Find x (a) 125 (b) 83 (c) 540 (d) 110

# VARIATIONS

*Variation can be described as the relationship between two quantities. Variation can also be called "proportion". It's a connection between quantities that described a relation connecting each other. Sit, relax and enjoy the FLAVOR OF MATHEMATICS.*

1.  **w varies directly to q and w is 2 and q is 3, Find the valves of q when w is 4.**  (a) 3    (b) 4    (c) 5   (d) 6

2.  **x varies directly to y and inversely as the square root of u. When x is 3, y is 4 and u is 16. Find the formula connecting all of them.**
   (a) $x = {}^{3y}/_{\sqrt{u}}$  (b) $x = {}^{3u}/_{\sqrt{y}}$ (c) $y = {}^{3u}/_{\sqrt{x}}$ (d) $y = {}^{3x}/_{\sqrt{u}}$

3.  **When M is inversely proportional to N, M = 16 and N = 12. Find the value of M when N = 24**
   (a) 4  (b)  8  (c) 12  (d) 14.

4.  **When $x = {}^{k}/_{\sqrt{y}}$, x = 7, y = 81 and k is constant, find x when y = 25**
   (a) 12.6   (b) 14    (c) 21.2   (d) 17

5.  **When y varies directly as the square root of w and inversely proportional to v; y = 7, w = 25 and v = 5. Find the valves of v when y = 6 and w = 4.**
   (a) ${}^{3}/_{7}$  (b) ${}^{3}/_{2}$  (c) ${}^{7}/_{4}$  (d) ${}^{7}/_{3}$

6.  **If V varies directly as P and V =15 and P = 4, find the formula connecting the later together.**
   (a) $V = {}^{P15}/_{4}$  (b) $P = {}^{V15}/_{4}$ (c) $V = {}^{154}/_{P}$ (d) ${}^{V}/_{15} = {}^{P}/_{5}$

7.  **N varies to M such that N = 16 and M = 2. Find the valve of M when N is 32**  (a) 1    (b) 2    (c) 3  (d) 4

8.  **A is directly proportional to B and inversely proportional to C. A = 8, B = 16 and C = 14. What is the valve of A when B = 12 and C = 6?**  (a) 12    (b) 13   (c) 14   (d) 15

9.  **The speed of a car is directly proportional to the time taken. If the speed is 20km and the time is 26hrs. Find the relationship of both the speed and the time taken.**
   (a) $S = {}^{T13}/_{10}$  (b) $S = {}^{T10}/_{13}$ (c) $S = {}^{10}/_{13t}$ (d) $S = {}^{13}/_{10t}$

10.  **If x varies to y and x = 2, y = 4; Find x when y = 6.**
   (a) 1    (b) 2    (c) 3    (d) 4

11.  **x varies to the square root of y; x = 4 and y = 64, find y when x = 8.**
   (a) 256    (b)  526    (c) 652  (d) 265

12.  **The volumes of a gas at constant pressure is inversely proportional to the temperature. When the volume is 35cm³, the temperature is 12. Find the temperature when the volume is 40cm³**
   (a) 11.8°  (b) 12.5°  (c) 10.5°    (d) 7.8°

13. The length of a plane shape called square is inversely proportional to the breadth. The length is 35cm and the breadth is 15cm. Find the breath when the length is 25cm.

(a) 19cm   (b) 20cm   (c) 21cm   (d) 22cm.

14. The diameter of a circle varies to the Circumference. The diameter is 5 and the circumference is 15. What is the diameter when the circumference is 9?   (a) 1   (b) 3   (c) 2   (d) 4

15. U is directly proportional to V and W is inversely proportional to Z. When U = 3, V = 4, W = 6 and Z = 72, Find the value of V and the formula connecting them When U = 7, W = 8 and Z = 28.

(a) $U = {}^{VWZ}/_a$, V = 27.8   (b) $U = {}^{VW9}/_z$, V = 2.72   (c) $U = {}^{VZ9}/_w$, V = 272   (d) $U = {}^{AW9}/_v$, V = 2

16. The electrical resistance of a wire is inversely proportional to the radius. If the resistance is 70Hz and Radius is 3cm, find the Radius when the Resistance is 40Hz.

(a) $R = {}^4/_{21}$cm   (b) $R = {}^2/_{21}$ cm   (c) $R = {}^{21}/_4$ cm   (d) $R = {}^4/_{21}$ cm

17. W varies directly as V and inversely as the cube root of y. W is 2, V is 4 and y is 64. Find the Value of V when W is 3 and y is 27.

(a) $V = 3^1/_2$   (b) $V = 4^1/_2$   (c) $V = 2^1/_2$   (d) $V = 1\frac{1}{2}$

18. N varies directly as S and inversely as the cube root of v. N is 4, S is 4 and y is 27. Find the Value of N when S is 3 and v is 64.

(a) $N = {}^9/_4$   (b) $N = {}^4/_9$   (c) $N = {}^3/_4$   (d) $N = {}^5/_7$

19. A Land that has to be cleared varies inversely as the square root of laborers that do the clearing. When 18 acre of land is cleared, 4 laborers are required. How many acre of Land did they clear having 9 laborers?

(a) 10   (b) 2   (c) 4   (d) 12

20. Z varies directly to $u^2$ and W is inversely proportional to v. Z = 3, U = 2, W = 1 and V = 4. Find z when U = 4, W = 2 and V = 3.   (a) 22   (b) 32   (c) 42   (d) 52

21. In a partial variation, a formula is given as B = V + ak. When B = 20, a = 2, and B = 40, a = 4. Find B When a = 6

(a) 40   (b) 50   (c) 60   (d) 70

22. The force E needed to make a machine full of load is partly constant and partly varies to the load to be pulled itself i.e (E = a + kL). When the load is 40kg, the force needed is 50N and when the load is 35kg, the force needed is 20N. The formula connecting them is

(a) E = – 190 + 6f   (b) E = – 190 + 40k   (c) E = – 50 + 240k
    (d) E = 196 + 20k

23. The heat developed in a wire varies jointly as the square root of the voltage applied and inversely to the resistance of the wire. When voltage is 225 volts, Resistance is 3 ohms, heat develop 50 calories per sec. What is the number of heat that will be developed when 49 volts are applied to a wire of Resistance 7 ohms?

(a) 33   (b) 10   (c) 2   (d) 20

24. The Illumination of a bulb varies inversely as the square of the distance. If the Illumination has 8 candle power at a distance of 2km, how many candle power does the illumination has on a bulb at a distance of 4km?    (a) 2    (b) 4    (c) 16    (d) 7

25. The mass that support the beam of a given thickness varies directly as the breath and inversely as the length. If the breath of 2cm, length 18m can support a mass of 100kg, find the mass which can be supported by a beam of 6cm broad and 40m long?
    (a) 120kg   (b) 130kg   (c) 145kg   (d) 135kg.

26. The time taken for a social meeting is partly constant and partly varies as the square of the number of member present. if there are 14 members present, the meeting will last only 45hrs, but with 22 members, it will last exactly 50hrs. How long will the meeting last if there are 19 members? (a) 8mins (b) 38mins (c) 28mins (d) 18mins

27. P Varies to QR. When Q = 9, R = 4, and P = 6. Find P When Q = 7 and R = 2.                (a) 3.2  (b) 1.2  (c) 3.5  (d) 2.3

28. y varies directly as x and inversely as the square of d. when y = 20, x = 16 and d = 8. Find y when x = 6, d = 14. (a) 2.4  (b) 3.9  (c) 5.5  (d) 7.8

29. The time of a pendulum varies as the cube root of its length. If the length of the pendulum that beats 12sec is 8cm. Find the law Connecting them.
    (a)$T = 4(\sqrt[3]{L})$   (b) $T = 6(\sqrt[3]{L})$   (c) $T = K\sqrt[3]{L}$ (5.9)   (d) $T = (2.3)\sqrt[3]{L}$

30. The number of spherical shot which can be cast from a given volume of lead varies inversely as the cube of the diameter of the shot. When the diameter is 3mm, the number of shot is 240. How many shot of diameter 2mm can be cast from the same volumes of lead?
    (a) 190    (b) 290    (c) 820    (d) 810.

31. The resistance of a car varies directly as the speed. If the Resistance is 605 and Speed is 40, find speed when Resistance is 500.
    (a) 29.2    (b) 33.1    (c) 35.2    (d) 38.7

32. The volume of a gas at constant pressure is inversely proportional to the temperature. If the Volume is $60m^3$ and temperature is 16Hz, find the volume, when the temperature is 8Hz. (a) $110m^3$ (b) $120m^3$ (c) $130m^3$ (d) $140m^3$

33. The speed of a toy van varies directly to the amount bought and inversely proportional to the voltage it acquires from an electric plug. Let's assume that the toy van speed is 45km in sec, voltage is 50V, amount bought is \$350. Find the formula connection
    (a) $S = {}^{\$V}/_{6.4}$     (b) $S = {}^{V6.4}/_{\$}$    (c) $S = {}^{\$(6.4)}/_{V}$    (d) $S = {}^{V}/_{\$6.4}$

34. The volume of a gaseous substance varies directly as the square of the pressure. When the gas is $512cm^3$, the pressure is 8Hz. Find the pressure when the temperature is $128cm^3$
    (a) 1 Hz    (b) 2Hz    (c) 4Hz    (d) 3Hz

35.    $1/\sqrt{r}$ varies to P and is inversely proportional to q. When r = 4, P = 2 and q = 8, find P when r =16 and q = 4   (a) $^2/_3$   (b) ½   (c) $^2/_5$ (d) ¾

36    P varies directly as the square of U and inversely proportional to the cube root of Z. When P = 9, Z = 1000 and U = 3, find P when U = 4 and Z = 64.          (a) 40     (b) 45     (c) 35     (d) 30

37.    The current that runs from the electric transformer to a live wire is inversely proportional to the current that runs from the live wire to every home circuit. if the current from the electric transformer is 68Hz and the other is 4. What is the current from the electric transformer when the current that runs from the live wire to every home circuit is 16?              (a) 16   (b) 17   (c)18   (d) 19

# ARITHMETIC PROBLEMS

*This is one of the major branch of mathematics. It deals with the act of expanding or converting word problems into equation to get a given answer that tally to the given question. Arithmetic problems deal with the act of imagining different aspect of calculations in our day to day activities and other given circumstances. I will advise you to be in full concentration and solve all questions in AGRESSION in this chapter. Enjoy the taste and the FLAVOR OF MATHEMATICS.*

1. **In a cinema hall, the ratio of women to men is 2:3, if there are 120 women, the number of men present in the hall is** (a) **120** (b) **140** (c) **180** (d) **200**

2. **Stanley, George and Isabel share $500 in the ratio of 4:6:12. How much did George get?** (a) $90.90 (b) $27.72 (c) $136.36 (d) $499.9

3. **Lawrence went to the market and realizes that 16 bags of pizza cost $120. If he buys 20bags of pizza, how much is the cost of pizza?** (a) $150 (b) $200 (c) $180 (d) $210

4. **It took 16 men to complete the building of the White House in 28 days. How many days will 8 men takes to complete the White house?** (a) 42 days (b) 52 days (c) 56 days (d) 62 days

5. **Lisa Scored 20 marks out of 80 marks. What is the percentage of her mark?** (a) 22% (b) 25% (c) 27% (d) 29%

6. **When a long ruler of length 20cm is measure by Chris to be 19.5cm, what is the percentage error?** (a) 2% (b) 2.5% (c) 3% (d) 3.5%

7. **Lambert and her sister share 30 oranges in the ratio of 2:3. How many oranges did each of them have together?**
   (a) Lambert = 18, sister = 12 (b) Lambert = 8, sister = 10.
   (c) Lambert = 12, Sister = 18 (d) Lambert = 10, Sister = 12

8. **Daniel and Peter share $147 between them so that Daniel get $19 more than peter. Find how much money each gets.**
   (a) Daniel = $83, Peter = $64 (b) Daniel = $64, Peter = $83
   (c) Daniel = $89, Peter = $58 (d) Daniel = $58, Peter = $89

9. **Mr. Franklin was traveling and he left a sum of the $30,000 for his family whereby $1/5^{th}$ was meant for the wife, $3/5^{th}$ was meant for his three sons and the remainder is to be shared among the two daughters. How much does each son gets in the amount?**
   (a) $4,000 (b) $6,000 (c) $9,000 (d) $3,000

10. **The age of Frank and Vicky are in the ratio of 2:5. If the age of Vicky is 15, find Frank's age three years ago.**
    (a) 4yrs (b) 3yrs (c) 5yrs (d) 6yrs

11. Fred and Frank were writing an English Examination. Fred obtained 56 marks more than Frank. If Fred obtain one – third of his own mark, he would have scored 12 marks more than twice Frank's mark. Find the mark scored by Fred.  (a) 40    (b) 50    (c) 60    (d) 70

12. The erection of the Amazon Towers takes 24 men to complete in 15 days. How many days will 9 men take to erect the tower when working at the same time?
(a) 30 days  (b) 40 days  (c) 50 days  (d) 60 days

13. A Network tower takes 26men to complete in 20days. How many days will 13 men use in completing the tower?
(a) 20 days    (b) 30 days    (c) 40 days  (d) 50 days

14. Dangote Limited produced 400 bags of cement a day. One Thursday, 350 bags were produced. What is the percentage decrease?
(a) 12. 5%  (b) 13.5%  (c) 14.3%  (d) 16.5%

15. Find the product of 12% of $200 and 9% of $400 when both of them are increased.                (a) 684        (b) 468    (c) 864        (d) 648

16. What is Mercy's selling price to make a profit of 22% when she sold a pack of wine for $450 and she made a profit of 20%?
(a) $40.66    (b) $406.67    (c) $4.067        (d) $406.7

17. Susan bought 600 oranges in which 42 of them were bad, 88 of them were bitter and the remaining was sweet. Find the difference between the percentage of the bad ones and the sweet ones.
(a) 10.2%  (b) 121. 3%  (c) 113.7%  (d) 71.3%

18. The length of a table measure as 16cm was measured by John as 15.5cm. What is the percentage error?                                    (a) 3.125%    (b) 31.25% (c) 312.5%    (d) 35.9%

19. There is a certain number whereby when it is divided by 15, it has a remainder of 5 and when the same number is divided by 22, It has a reminder of 19. Find that number
(a) 695   (b) 569   (c) 659   (d) 965

20. Brenda and her husband went to a supermarket to buy provisions which cost $2000. Brenda had 20% of the cost and her husband had 60% of the remainder. How much did they have altogether?
(a) 1420  (b) 3106  (c) 1360  (d)    603

21. Emmanuel spends ¾ of his salary on rent, sales and food. He spent $1/_{10}$ of the remainder on shelter. What percentage of his salary is left?
(a) $22^1/_2\%$    (b) $20^2/_3\%$    (c) $19^1/_2\%$    (d) $21^2/_3\%$

22. When Rachel and Daniel went to a company to work, Rachel's wages was given at $120 per day and Daniel's wages was $150 per day. The difference in their wages is......

(a) \$150   (b) \$20   (c) \$270   (d) \$30

23. A long wood of 400cm long is cut to four pieces in the ratio of 1:2:3:4. Find the length of the third wood.
(a) 40cm   (b) 80cm   (c) 120cm   (d) 160cm

24. The mean of 6 positive numbers is 20. When another number is added, the mean becomes 25. Find the seventh number.
(a) 40   (b) 45   (c) 50   (d) 55

25. When Samuel is working in Indiana Petroleum Limited, his annual increment was 20% of his present salary which is \$30,000. What is his annual salary at the beginning of the 4th year?
(a) \$39,600   (b) \$36,000   (c) \$43,560   (d) \$45,595

26. What is the difference between $1/4$ of a certain number and $1/5$ of the same number when the result equals to 4?
(a) 50   (b) 60   (c) 70   (d) 80

27. If John bought a car at \$180,000 and sold it at \$120,000. What is the loss percentage?   (a) 23.3%   (b) 33.3%   (c) 43.3%   (d) 50%

28. Mr. Welkin and his brother invested a total of \$5,000 on an irrigation project. The farm yield was sold for \$15,000 at the end of the season. If the profit was shared in the ratio 2:3, what is the difference in the amount of profit received by both of them?
(a) \$1,000   (b) \$2,000   (c) \$3,000   (d) \$4,000

29. The Angel investors invested a sum of \$50,000 in two companies. If these companies pay dividend of 6% and 8% respectively, how much did they invest at 8% if the total yield is \$3,700?
(a) \$25,000   (b) \$30,000   (c) \$35,000   (d) \$40,000

30. Two numbers differ by 8, their product is 65. Find the numbers
(a) 10 and 8   (b) 13 and 5   (c) 19 and 11   (d) 12 and 4.

31. Find the positive number x such that the cube of the number is equal to nine times the number.   (a) 1   (b) 2   (c) 3   (d) 4

32. Mr. Welford is 28years older than his daughter. In 2 years' time, he will be 3 times as old as his daughter. Find their present ages.
(a) Welford = 50, Daughter = 30   (b) Welford = 40, Daughter = 12
(c) Welford = 48, Daughter = 22   (d) Welford = 42, Daughter = 15

33. A certain number is added to 9 and 8 is subtracted from that same number. Their product is given as 0. What is the number?   (a) 9   (b) 8   (c) 7   (d) 6.

34. What is the difference between one – fourth of 128 and half of 42? (to 2 s.f)   (a) 43   (b) 32   (c) 11   (d) 21

35. James subtract 3 from the numerator of a fraction and Kurt Subtract 4 from the denominator of the same fraction and it gives $3/7$. What is the

fraction?
    (a) $^9/_4$    (b) $^{28}/_{33}$    (c) $^{28}/_{31}$    (d) $^{31}/_{28}$

36. When one – half of a number is added to two – third, it is equals to one – fourth of that same number subtracted from one – seventh.  Find that number.    (a) $2^7/_9$  (b) $-3^8/_{19}$  (c) $-^{44}/_{147}$  (d) $4^2/_{17}$

37. A trader marked his goods to have a profit of 25%. He allows 20% discount for cash. Find his percentage profit when sold for cash.
(a) 1% (b)    2%    (c)3%  (d) 4%

38. A Radio set is sold for \$1225 to gain $22^1/_2$%. What is the actual gain?
(a) \$150 (b) \$200 (c) \$ 225 (d) \$275

39. An agent receives 5% commission for transacting a sales business worth \$20,000 for his principal. Calculate the commission involved in the amount.
(a) \$4000    (b) \$1000    (c) \$2000    (d) \$3000

40. Two partners Benny and Lawson invest \$3000 and \$1800 respectively in a business. It is agreed that Lawson should take 30% of the profit for running the business and that remaining profit should be divided between them in the ratio of the capital invested. Find what percentages of the profits benny receives.
(a) $46^3/_4$%    (b) $38^4/_9$%    (c) $53^4/_9$%    (d) $43^3/_4$%

41. Marriage duties of 20% are paid on a legacy of \$4,500. The eldest son takes 50%, the second son 30% and the youngest took the remainder. What percentage of the original legacy does the younger son receive?
(a) 16% (b) 18%    (c) 20% (d) 22%

42. Find the principal which amounts to \$2000 in 5 years at 2% simple interest.    (a) \$1,000 (b) \$1,581 (c) \$1,818 (d) \$2,818

43. A sales man receives a commission 4% of the value of the goods he sells. Find the value of the goods he sold when his commission is \$62.
(a) \$1,550    (b) \$2,152    (c) \$2,550  (d) \$1,570

44. The simple interest of \$295 at 3% for 7 years is
(a) \$95.61    (b) \$61.95    (c) \$59.16    (d) \$16.59

45. A string was actually 4.8m. David measured it to be 4.95m. Find the percentage error.    (a) $3^1/_{33}$%    (b) $3^1/_8$% (c) $1^5/_{16}$%  (d) 15%

46. Mr. Stephen earns a Salary of \$50,000 p.a with three dependent relatives of up to maximum of two (\$2,400) with seven children (for up to maximum of three for \$2,000) and pays \$5,000 to social schemes, \$5,000 for tithe in church. How much did he pay for tax, in which his personal allowance included is \$5,000?
(a) \$4,600    (b) \$5,600    (c) \$5,670    (d) \$5,630

47. Half of a certain number is seven more than three – fourth. What is the number?    (a) $^{12}/_{13}$  (b) $^{31}/_2$  (c) $^2/_{31}$  (d) $^{16}/_{17}$

48. In order for Mr. Clarkson to pay his daughter's School fees, he Invest $5,000 at 4% p.a. if the money is invested at the beginning of the girl's school career and $1,800 is withdrawn at the end of each year for the fees, find how much is left after 3 years
(a) $941.90   (b) $900.40   (c) $928.42   (d) $950.20

49. Mandela buys a house for £25,000 and after allowing an annual renovation of £400 wishes to make a 5% p.a return on his money. What should be the annual rent of the house?
(a) £1,700       (b) £1,655       (c) £1,650       (d) £1,600

50. Find the simple interest on a loan of $2,000 for 4years at 5% per annum.
(a) $200  (b) $ 300  (c) $250  (d) $400

51. If $210 amount to $235.20 at 3% p.a. Find the time in simple interest.
(a) 4 years (b) 3 years  (c) 2 years (d) 1 year.

52. Find the compound interest when David burrows $2,500 at the rates of 6% p.a in 2 years.    (a) $300  (b) $309  (c) $350  (d) $395.

53. A loan of $6,000 generates a simple interest of $1,350 for $7\frac{1}{2}$ years. Find the rate per cent        (a) 1%  (b)  2%  (c) 3%        (d) 4%

54. Jackson paid $500 for a textbook over which he enjoyed 10% cash discount. How much was the marked price?
(a) $500              (b) $556         (c) $596          (d) $692

55. If there are 225 students in a school and 150 of them are girls; what is $1/5^{th}$ of the number of the boys? (a) 15   (b) 75    (c) 120    (d) 60

56. The chairman of Globacom communication earn a salary of $90,000.00 His personal allowance is $5,000, he is married with three children in which their allowance is $2,000 and donate to the mother less babies home in allowance of $10,000. Find how much he pay for tax.
(a) $20,500   (b) $22,176   (c) $36,070   (d) $42,078

57. How much commission will an agent receive from the sales of $250 at the same rate of 15cent?   (a) $38  (b) $42  (c) $45.9  (d) $37.5

58. Find the compound interest on $950 in 6 years at 3% p.a.
(a) $184.3       (b) $162.9   (c) $140.2   (d) $180

59. Mr Eric burrows $600 from a co-operation at 3% p.a interest. If he repays $300 at end of each year, how much does he still owe after 2nd repayment?
(a) $120  (b) $127.54  (c) $130  (d) $172.28

60. Aaron, Clarion and Carolina Share the sum of $1,495. Aaron's share is ¾ of Carolina's share and Clarion's share is $5/7$ of Carolina's share. Find Aaron's share.        (a) $400       (b) $455  (c) $356          (d) $500

61. Two – third is multiplied by the sum of a certain number and 5 while one – sixth is multiplied by the sum of the same number and 2. If the sum of the two expressions equals to 7. What is that number?
(a) 7   (b) 6   (c) 5   (d) 4

62. David and stone share $10 in the ratio of 2:4. What is the difference between the shares of the both of them?
(a) 6.67   (b) 3.33   (c) 3.34   (d) 4.23

63. Jude bought a shoe that cost $370. The shoe was sold at a loss of 14%. What is the selling price? (a) $450   (b) $318.20   (c) $51.8   (d) $340.7

64. A house bought for $100,000 was later auctioned for $80,000. Find the loss percent. (a) 40% (b) 30%   (c) 50%   (d) 20%

65. Mrs. Jackson sold an article for $7.50 instead of $12.75. Calculate the percentage error, correct to one decimal place.
(a) 1.7%   (b) 41.2%   (c) 18.3%   (d) 5.3%

66. Steven gave $5,852 to Gerald, Joe and Kingston so that Gerald's share is one fourth of Joe's share and Joe's share is two – third of kingston's share. What is the sum of Gerald's share and Joe's share? (a) $2,600   (b) $2,900   (c) $2,660 (d) $2,000

67. Kingsley and Kingston received $190 to share it among themselves. If Kingsley received $20 more than Kingston, how much did Kingston received? (a) $70   (b) $75   (c) $105   (d) $190

68. 8 is subtracted from the product of a certain number and ¾ while 7 is added to $^1/_3$ of the same number. The difference between the two expression is 12. What is the number?
(a) 40   (b) 30   (c) 64   (d) 46

69. Bill spent ¼ of his money on shelter and $^2/_3$ of the same money on transport. She had $25 left. How much does she had at first?
(a) $500   (b) $600   (c) $400   (d) $300

70. When there are 333 students in a classroom, $^2/_9$ of them are girls. How many boys are there?
(a) 200 boys   (b) 259 boys   (c) 278 boys   (d) 299 boys

71. Adam spent $^3/_7$ of his money on shoes and $^1/_5$ of his money on cloth. What was the amount he had at first if the remainder of the money is $130?   (a) $400   (b) $350   (c) $300   (d) $250.

72. The mean of eleven positive numbers is fifteen. When another number is added, the mean becomes Nineteen. Find the twelfth number. (a) 52 (b) 57   (c) 63   (d) 72

73. A certain number is divided by 33. It gives 4 and has a remainder of 16, the same number is also divided by 28, it gives 5 and has a remainder of 8. What is that number?
      (a) 132     (b) 138  (c) 148  (d) 153

74. What is that number whereby the cube of four is equal to the square of 8?      (a) 32   (b) 64  (c) 132  (d) 140

75. Daniel has $x and Bill has half of what Daniel have. If $20 is subtracted from what Daniel have and $40 is added to what bill have, how much does bill have?
      (a) $60   (b) $68   (c) $72   (d) $120

76. Joshua scored 75 marks in English instead of 57, his average mark in 4 subjects would have been 60. What was has total mark?
      (a) 122   (b) 182  (c) 212  (d) 222

77. A man gave Terry $x and Teddy $45. Terry realizes that the money given to him was too small and he forcefully took $1/5$ of Teddy's money. How much does terry has with his $x?
   (a) $(x + 6)  (b) $(x + 9)  (c) $(x – 9)  (d) $36

78. Jack accused Rose of taking $x from his $900 without his consent. Rose decided to give him $30 more to his money in which everything equals $145. What is jack's total amount?
      (a) $760  (b) $785  (c) $800  (d) $845

79. Frank has $x and Michael has $5 plus twice of what Frank has. Frank realizes that $45 is added to his money. How much Michael did has?
   (a) $30  (b) $35  (c) $40  (d) $85

80. Bill gate and Carlos Slim Share the sum of $20,000,000. If Bill Gate get twice of $4,000,000, How much is Carlos Slim's share?
   (a) $2,000,000   (b) $8,000,000   (c) $12,000,000  (d)$10,000,000

# SET THEORY

*A Set can be described as the grouping or collection of objects or elements. There are lots of terms governing the rules of set and students solving this chapter must be familiar with the rules. Sit, relax and concentrate on the FLAVOR OF MATHEMATICS..*

1.  If z = {u , b, h, b, n, q}. Find ∩ (z)   (a) 2  (b) 3  (c) 4  (d) 5

2.  **Which of the following set are disjoint?**
    (a) {1, 3, 4} & {5, 6, 9}  (b) {1, 2, 3} & {1, 2, 3} (c) {2, 1, 4} & {4, 1, 2}        (d) {9, 2, 5} & {5, 9, 2}

3.  **If X = {2, 3, 4} and Y = {1, 3, 5}, what is X ∪ Y?**
    (a) {3, 4, 5}   (b) {1, 2, 3, 4, 5}    (c) {3, 2 , 1 , 4}   (d) {1, 3, 4, 5}

4.  **The power set of V = {1, 2, 3, 4} is**
    (a) 12 subset  (b) 14 Subset  (c) 16 Subset  (d) 8 Subset

5.  **When x = {a, b, c, d} and V = {d, a, b, c}, this set is said to be**
    (a) Equal      (b) Infinite  (c) Finite   (d) Disjoint

6.  Find ∩ (Q) when Q is {1, 3, 5, 7, 9}.   (a) 2      (b) 3    (c) 4    (d) 5

7.  **If the Universal set U = {a, b, c, d, e, f, g, h, i} x = {a, b, e, f, g, h}, y = {a, b, c, e, g, i} & {x¹ n y}**
    (a) {a, b, e, q}   (b) {c, i}  (c) {b, c, g, i}  (d) {a, b, c, d}

8.  **If x = {1, 2, 3, 4, 8, 9, 10}, y = {1, 3, 4, 9} and z = {1, 2, 3, 8, 9, 10}. Find {x ∪ y ∩ z}**
    (a) {1, 2, 3, 8, 9, 10}   (b) {1, 2, 3, 4, 8, 9, 10}   (c) {2, 3, 8, 10}  (d) {1, 2, 8, 9, 10}

9.  **When a universal set U = {a, b, c, d, x, f, g, i} A = {b, x, f, i} and B = {a, b, f, i , d}. Find A¹ ∩ B**
    (a) {a, d}    (b) {b, f, i}     (c) {x, f, d}

10. **When X = {1, 3, 7, 9}, Y = {3, 7, 2, 4, 17, 19} Z = {1, 3, 4, 7, 8, 11}. Find X ∪ Y ∩ Z**
    (a) {1, 4, 7}    (b) {1, 3, 4}          (c) {1, 3, 4, 7} (d) {1, 7}

11. **When a universal set B = {a, b, c, d, e, g, h, u, m, n} x = {a, e, h, n, g} and y = {c, b, d, g, m}. Find {x¹∩ y¹}**
    (a)  {g}         (b)  {u}    (c) Φ   (d) {e, h, g}

12. **P = {Prime numbers less than 30}, Q = {Odd numbers less than 12} R = {3, 7, 11, 15, 17}; Find P ∪ Q ∩ R.**
    (a) {3, 11}    (b) {3, 9, 11, 17}    (c) {3, 7, 11, 17}   (d) {3, 7, 9, 11}

13. The Universal is given as U = {i, n, t, e, r, e, s, u}, V = {t, e, i, n, u} & W = {e, t, s, n}.          Find {v $\cup$ w$^1$}
    (a) {e}    (b) {u}   (c)  {t, e, i, n, u, r}   (d) {e t, n}

14. When v is a universal of V = {x: x $\leq$ 9}
    U = {x + 1, x < 4}, W= {1, 4, 6, 7, 8}. Find {W$^1$ $\cup$ U} $\cap$ W
    (a) {1}   (b) {2, 3, 4}   (c) $\Phi$          (d) {4}

15. When x is given as {even number less than eleven} is
    (a) {2, 4, 6, 8, 9}   (b) {2, 4, 6, 8, 10 }   (c) {2, 6, 8, 10}   (d) {2, 4, 6, 8}

16. When Y = {b, d, e, g, h} and C = {u, v, w}. Then Y $\cap$ C is
    (a) {b, e, f }(b) {u, v, h, b, e, f}   (c) { b, d, f, c}   (d) $\Phi$

17. When X = {a, b, c, d, e, f} and Y = {b, d, e, g, h}.  Find Y – X
    (a)  {g, h}    (b)  {g, d, e, f}    (c) {d, f}    (d) {f}

18. When A = {1, 2, 4, 6, 8, 10} and B = {4, 6, 8, 9}
    Find A $\cap$ B.        (a) {1, 2, 10}   (b) {4, 6, 8}    (c) {1, 2, 9, 10}    (d) {1, 9}

19. A circle diagram that shows the relationship between sets is known as ........
    (a) circle venn   (b) venn diagram   (c) kite venn   (d) Venn theory

20. If K = {x: -i $\leq$ x < 5 }, Y ={ x : x $\leq$ 4}, Z = {-1, 2, 4, 6}. Which numbers are common in the three given sets?
    (a) {0, 2, 4}      (b) {-1, 4}      (c) {-1, 3, 6}      (d) {2, 4}

21. In a girls group, there are 190 girls. 90 wear blue uniform, 68 wear black uniform and 120 wear green uniform. If 20 wear blue and black uniform, 42 wear green and black uniform, 38 wear green and blue uniform, the number of girls that wear the three uniforms is
    (a) 10    (b) 11    (c) 12    (d) 13

22. {A $\cap$ B} $\cup$ {A $\cap$ C} is the same as
    (a) A$\cap$ (B $\cup$ C}   (b) {A $\cap$ B} C   (c) {A $\cap$ B} $\cap$ {A $\cap$ C}   (d) A $\cap$ {B $\cup$ C}

23. In a foreign language school, there are 350 students. 170 study French, 190 study German and 180 study Spanish. If 70 study French and German, 60 study French and Spanish, 80 Study German and Spanish and 30 students study the three languages. What is the number of student that study only one language?          (a) 160    (b) 170    (c) 180    (d) 210

24. V = {Odd numbers less than 20}, Q = {Prime number less than 20}
    W = {1, 5, 13, 17, 19}. Find V $\cap$ Q $\cap$ W.
    (a) {1, 5, 13, 17, 19}   (b) {5, 17, 19}   (c) {5, 13, 19}   (d) {5}.

25. A class contains 26 students, whereby 16 of them use ink while 10 of

them use pencil. If 4 of them do not use either ink or pencil, find the number of student that uses both ink and pencil.
(a) 3 Students  (b) 4 Students   (c) 5 Students  (d) 6 Students

26. In a Class, there are 480 Students whereby 210 students use Calculator, 350 students use their Ipad and 300 students uses Laptops. 90 students uses Calculator and Laptops, 110 students uses their Ipad and Calculator and 190 students uses their Ipad and Laptops. If 10 students use the three devices, find the number of students that uses only two devices. (a) 310    (b) 320    (c) 360    (d) 390

27. What can also be called an empty set in the set theory?
(a) Set notation  (b) Complement      (c) Universal set   (d) Null set

28. When P = {1, 2, 3, 4} and Q = {5, 8, 9, 11}.The Intersect of the two set is
(a) {2, 3}    (b) {1, 2, 3, 4, 5, 8, 9, 11}    (c) $\Phi$    (d) {5, 2, 9, 11}

29. In the Brooklyn Game reserved, there are 400 animals. 240 animals are dogs, 250 animals are cats and 230 animals are snakes. If 110 animals are cats and dogs, 100 are snakes and cats, 120 are dogs and snakes. The number of animals that include the dogs, cats and snakes are
(a) 8     (b)  9    (c) 10          (d) 11

30. The element of the set of P and K below is
      P = {Factors of 20} and K= {Multiples of 3 less than 12}
(a) P= {2, 4, 5, 10} and k = {3, 6, 9}   (b) P = {2, 4, 10} and K = {6, 9}
(c) P= {2, 10} and K = {$\Phi$}               (d) P = {$\Phi$} and K= {3, 6}

31. The element of set U and T is U= {Prime numbers less than 30}, T = {Even numbers less than 13}, what is U and T?
(a) U = {2, 3, 7, 9, 10}, T = {2, 4, 6, 10}
(b) U= {2, 3, 5, 7, 11, 13}, T = {2, 4, 6, 8, 10, 12}
(c) U = {2, 3, 5, 7, 11, 13, 17, 19, 23, 29}, T = {2, 4, 6, 8, 10 12}
(d) U = {1, 2, 3}, T = {$\Phi$}

32. When T = {1, 2, 3, 4, 5, 6}, L = {2, 1, 4, 6}, K = {2, 3, 7}, Find K $\cup$ L $\cap$ T
(a) {2, 3, 6}   (b) {1, 2, 3, 6}   (c) {2, 3, 4, 6}   (d) {1, 2, 3, 4, 6}

33. When K = {2, 3, 7} and L = {2, 1, 4, 6}, Find K $\cap$ L
      (a)  {2}     (b) {2, 3}     (c) {1, 4, 7}     (d) {2, 1, 4, 6}

      When a Universal set U = {1, 4, 6, 8, 5, 9, 11, 13}
   T = {4, 9, 13, 11}, K = {6, 9, 13} and L = {1, 8, 5, 9}

34. Find L' $\cup$ T'$\cap$ K   (a) {5, 6, 9, 13}   (b)  {1, 5, 8, 9}  (c)  {6, 13}  (d) $\Phi$

35. Find K' $\cup$ T' $\cup$ L'
   (a) {1, 4, 6, 8, 5, 11, 13}  (b) {1, 4, 6, 13}  (c)  $\Phi$ (d) {1, 4, 6, 5, 8, 11, 13, 15}

36. Find K $\cap$ T$\cap$ L'          (a) {13}   (b) {9}    (c) $\Phi$   (d) {1, 6, 11, 13}

37.  If x = {a, b, c, d, e, f, g}, u = {a, d, e, i, g}, f = {d, c, g, i, u}. Find (f ∩ u ∩ x)
     (a) Φ      (b)  {d, g}    (c)     {a, d, l}        (d) {a, d, r, e}

38.  The cardinal number of the set is U = { A: a E A,10 > a}
     (a) {1, 2, 5, 7, 9}    (b) {1, 2, 3, 4, 5, 6, 7, 8, 9}   (c) {1, 2, 5, 7, 8, 9}        (d)
{1, 3, 5, 8, 9}

39.  The set A = {Positive integers whose cubes are less than 50} is given
     as    (a) {2, 3}    (b) {1, 2, 3}    (c) {2}    (d) {3}

40.  Set B = {x : x E even numbers from 10 to 20} is given as;
     (a) {12, 14, 18}      (b) {12, 14, 16, 18}    (c) {12, 18}    (d) Φ

41.  Which of the options below is set G = {Prime factors less than 20}?
     (a) {2, 3, 5, 7} (b) {2, 3, 5, 7, 11, 13, 17, 19}  (c) {2, 5, 7, 9, 17}  (d) Φ

42.  Which set options below represent the set given E = {All number less
     than 10 which are not multiplies of 2}
     (a)  {1, 3, 7, 9}          (b)  {1, 7, 9}      (c) {1, 3, 5, 7, 9}     (d) {3, 7, 9}

     When U = {2, 4, 5, 8, 9, 11, 12, 15, 28, 30, 32}
     A = {4, 9, 11, 12, 15}, B = {2, 5, 9, 30, 32} and C = {2, 5, 8, 9, 11}

43.  Find A' ∪ C'
     (a) {2, 4, 5, 12, 8, 15, 28} (b) {2, 4, 5, 15, 28}  (c) Φ
     (d) {2, 4, 5, 8, 12, 15, 28, 30, 32}

44.  Find B' ∩ C' ∩ A    (a) {4, 12, 15}   (b) {4, 15}        (c) {4, 12} (d) {12, 15}

45.  A = {x, 3x, 4x, 9x, 5x, 8x} and B= {3x, 4x, x, 8x}, Find A ∪ B
     (a) {x, 4x, 8x, 9x}   (b) {x, 4x, 5x, 8x}  (c) {x, 3x, 4x, 5x, 8x, 9x}
     (d) {x, 3x, 4x}

46.  There are 30pupils in a language school learning French and Chinese
     language. If 15 pupils learn French, 10pupils learn Chinese and 7pupils
     learn both languages together. How many pupils learn ONLY French
     language?  (a) 5  (b) 3    (c) 4     (d) 8

47.  In a hall in which 60 people entered; 28people can speak French,
     32people can speak Chinese, 26people can speak Portuguese, if
     10people can speak French and Chinese, 8poeple can speak French and
     Portuguese, 12 people can speak Chinese and Portuguese and also at
     the same time, 4 people can speak the three languages. Find the
     numbers of people that can speak two languages.
     (a) 25 people  (b) 18 people  (c) 35 people (d) 40 people

48.  It was recorded that a total of 83 people read New York Times, 27 read
     Forbes news, and 42 read Newsy. If 7 people read Forbes news and New
     York Times, 10 people read Forbes news and Newsy, 12 people read the
     New York Times and Newsy, 5 people read the three newspapers, Find
     the number of people that read one newspaper.
     (a)  109 people (b) 72 people (c) 84 people (d) 104 people.

49. In an Exam hall, 20 students offer biology, 30 students offer chemistry, 25 offer physics. In this exam hall, 12 offer biology and physics, 8 offer physics & chemistry 4 offer both the three subjects. How many students were found in the hall if 10 offer biology and chemistry?
(a) 42 People (b) 44 People (c) 47 People (d) 49 People

50. 134 Boys went into the forest to cut down trees. 70boys cut down Mahogany, 56boys cut down Ebony and 50boys cut down Blackwood. 19boys cut down Blackwood and Ebony, 21boys cut down Mahogany and Blackwood, 18boys cut down Mahogany and Ebony. Find the number of boys that cut down the three types of tree.
(a) 14 boys (b) 16 boys (c) 20 boys (d) 22 boys

51. In an airport, 25 people were going to London, 35 people were going to USA, 29 People were going to Nigeria. if 18 people were going to London and Nigeria, 19 people were going to USA and Nigeria, 15 people were going to London and USA and 12 people are going to the three countries, Find the number of people that went for the flight.
(a) 49 people (b) 53 people (c) 57 people (d) 64 people

If U is a universal set $\{1 - 10\}$
$x = \{2, 4, 6, 8\}$, $y = \{1, 3, 5, 7\}$ and $z = \{3, 4, 6, 9\}$

52. What is $X \cap Y$?  (a) $\{2, 6\}$ (b) $\Phi$ (c) $\{3, 5, 7\}$ (d) $\{1, 3, 6\}$

53. What is $X' \cup Z$?
(a) $\{1, 3, 5, 7, 9, 10\}$ (b) $\{1, 3, 4, 5, 6, 7, 9, 10\}$ (c) $\{3, 5, 9, 10\}$
(d) $\{1, 7, 10\}$

54. What is $X' \cap Z$?
(a) $\{3, 4, 6, 9\}$ (b) $\{1, 3, 5, 7, 9, 10\}$ (c) $\{3, 9\}$ (d) $\{1, 4, 5, 9\}$

55. What is $Z' \cap Y$?
(a) $\{1, 2, 5, 7, 8, 10\}$ (b) $\{2, 8, 10\}$ (c) $\Phi$ (d) $\{1, 5, 7\}$

# LOGARITHMS

| | | | | | | | | | |
|---|---|---|---|---|---|---|---|---|---|
| 1. B | 7. C | 13. C | 19. D | 25. A | 31. B | 37. B | 43. A | 49. A | 55. A |
| 2. C | 8. A | 14. D | 20. A | 26. B | 32. B | 38. D | 44. B | 50. A | 56. C |
| 3. A | 9. C | 15. B | 21. D | 27. C | 33. B | 39. D | 45. A | 51. D | 57. B |
| 4. A | 10. C | 16. A | 22. A | 28. C | 34. B | 40. B | 46. B | 52. B | 58. C |
| 5. B | 11. A | 17. C | 23. B | 29. D | 35. C | 41. C | 47. D | 53. B | 59. C |
| 6. D | 12. D | 18. A | 24. B | 30. A | 36. B | 42. D | 48. C | 54. B | 60. C |

# ARITHEMETIC AND GEOMETRIC PROGRESSION

| | | | | | | | | |
|---|---|---|---|---|---|---|---|---|
| 1. D | 7. D | 13. A | 19. A | 25. D | 31. C | 37. B | 43. C | 49. C |
| 2. B | 8. B | 14. B | 20. B | 26. C | 32. D | 38. B | 44. A | 50. B |
| 3. B | 9. D | 15. C | 21. C | 27. B | 33. C | 39. A | 45. B | |
| 4. D | 10. B | 16. D | 22. A | 28. A | 34. C | 40. B | 46. A | |
| 5. D | 11. B | 17. C | 23. C | 29. B | 35. C | 41. C | 47. A | |
| 6. B | 12. C | 18. B | 24. A | 30. B | 36. A | 42. B | 48. C | |

# POLYGONS

| | | | | | |
|---|---|---|---|---|---|
| 1. C | 7. C | 13. C | 19. B | 25. B | 31. C |
| 2. B | 8. C | 14. C | 20. A | 26. C | 32. B |
| 3. C | 9. A | 15. C | 21. A | 27. B | 33. A |
| 4. B | 10. B | 16. A | 22. D | 28. A | 34. B |
| 5. C | 11. D | 17. B | 23. C | 29. B | 35. B |
| 6. A | 12. B | 18. D | 24. A | 30. A | |

## VARIATIONS

| | | | | | | |
|---|---|---|---|---|---|---|
| 1. D | 7. D | 13. C | 19. D | 25. D | 31. B | 37. B |
| 2. A | 8. C | 14. C | 20. B | 26. A | 32. B | |
| 3. B | 9. B | 15. B | 21. C | 27. D | 33. C | |
| 4. A | 10. B | 16. C | 22. A | 28. A | 34. C | |
| 5. D | 11. A | 17. B | 23. B | 29. B | 35. B | |
| 6. A | 12. C | 18. A | 24. A | 30. D | 36. A | |

# ARITHEMETICAL PROBLEMS

1. C   8. A   15. C   22. D   29. C   36. C   43. A   50. D   57. D   64. D   71. B   78. B
2. C   9. B   16. B   23. C   30. B   37. C   44. B   51. A   58. A   65. B   72. C   79. D
3. A   10. B   17. D   24. D   31. C   38. A   45. B   52. B   59. B   66. C   73. C   80. C
4. C   11. C   18. A   25. C   32. B   39. B   46. D   53. C   60. B   67. B   74. B
5. B   12. B   19. D   26. D   33. B   40. D   47. B   54. B   61. D   68. C   75. A
6. B   13. C   20. C   27. B   34. C   41. A   48. A   55. A   62. C   69. D   76. D
7. C   14. A   21. A   28. B   35. A   42. C   49. C   56. B   63. B   70. B   77. B

# SET THEORY

| | | | | | | | | | |
|---|---|---|---|---|---|---|---|---|---|
| 1. D | 7. B | 13. C | 19. B | 25. B | 31. C | 37. B | 43. D | 49. D | 55. D |
| 2. A | 8. A | 14. A | 20. A | 26. C | 32. D | 38. C | 44. A | 50. B | |
| 3. B | 9. A | 15. B | 21. C | 27. D | 33. A | 39. B | 45. C | 51. A | |
| 4. C | 10. C | 16. D | 22. D | 28. C | 34. C | 40. B | 46. D | 52. B | |
| 5. A | 11. B | 17. A | 23. D | 29. C | 35. A | 41. B | 47. B | 53. B | |
| 6. D | 12. C | 18. B | 24. A | 30. A | 36. A | 42. C | 48. A | 54. C | |

www.ingramcontent.com/pod-product-compliance
Lightning Source LLC
Chambersburg PA
CBHW082010160726
47999CB00008B/2772